COMPETITIVE BIOLOGY

INTRODUCTION

This objective biology series provides a basic and challenging problem of biology from particular topics. It can be used to brush up ones basics and checking up the preparation level of particular topic. It is equally helpful to the traditional classes as well as competitions. It can be also taken as a revision material for any competition which includes the test of basic biology. If you want to grasp the subject before practicing these multiple choice questions, you can go through the website http://www.ncert.nic.in/ncerts/textbook/textbook.htm and down load the free copy of science books and after having command on the topic practice it. For revision purpose, important points are given at the starting of each topic.

If you have any query or suggestion about this series you can send your suggestion at uk2594@gmail.com.

CONTENTS

1. CELL

SOME IMPORTANT POINTS

➢ Robert hooked discover the cell in 1665 with the help of a primitive microscope.

➢ Cell is the fundamental unit of life.

➢ The plasma membrane allows the entry or exit of some material in and out from the cell.

➢ The movement of substances from a region of high concentration to the region low concentration is known as differs.

➢ The movement of water from a region of high concentration to the low concentration is known as osmosis.

➢ Plasma membrane is flexible and is made up of organic molecules called liquids and proteins.

➢ Amoeba gets its food from the external environment by the process of endnocytosis.

➢ Cellulose is a complex substance and provides structural strength to plants.

➢ Chromosomes contain information for inheritance of features from parents to next generation in the form of DNA (Deoxyribo Nucleic Acid).

➢ Mitochondria is known as the power house of cell. Because the energy needed for some activities in stored in it is the form of ATP.

➢ Plants cells have cell wall while animal cells have not cell wall.

➢ Lysosomes are known as the suicide bags of cell.

➢ Plastids are only present in plant cell.

➢ Plant cells have big vacuoles in compression to animal cell.

➢ The primary function of leucoplasts is storage.

1. CELL

1.   The largest cell in the human body is?

a) Nerve cell   b) Muscle cell

c) Liver cell   d) Kidney cell

2.   The barrier between the protoplasm and the other environment in an animal   cell is?

a) Cell wall   b) Nuclear membrane

c) Neuron   d) Plasma membrane

3.   A plant cell differs from an animal cell in the absence of?

a) Endoplasmic   b) Mitochondria

c) Ribosome   d) Centrioles

4.   Centrosome is found in?

a) Nucleus   b) Cytoplasm

c) Chromosomes   d) Nucleus

5.   The power house of a cell is?

a) Chloroplast   b) Golgi apparatus

c) Nucleus   d) Mitochondria

6.   Within a cell the site of respiration is the?

a) Ribosome   b) Golgi apparatus

c) Endoplasmic reticulum   d) Mitochondria

7.   Which is called suicidal bag of cell?

a) Centrosome   b) Lysosome

c) Mesosome   d) Chromosome

8.   Ribosome is the centre for?

a) Protien synthesis   b) Respiration

c) Fat synthesis   d) None of these

9. Double membrane is absent in?

a) Mitochondria   b) Chloroplast

c) Lysosome   d) Nucleus

10. Cell organelle found only in plant is?

a) Golgi apparatus   b) Mitochondria

c) Plastids   d) Ribosome

11. Organisms lacking nucleus and membrane bound organelle are?

a) Diploids   b) Haploids

c) Eukaryotes   d) Prokaryotes

12. Animal cell is limited by?

a) Basement membrane   b) Plasma Membrane

c) Cell Wall   d) Shell Membrane

13. The Network of Endoplasmic Reticulum is present in the?

a) Nucleus   b) Nucleolus   c) cytoplasm   d) Chromosomes

14. Lysosomes are reservoirs of?

a) Fat   b) RNA

c) Secretary Glycoprotein   d) Hydrolytic Enzymes

15. The membrane surrounding the vacuole of a plant cell is called?

a) Tonoplast   b) Plasma Membrane

c) Nuclear Membrane   d) cell wall

16. Cell Secretion is done by?

a) Plastids   b) ER   c) Golgi   d) Nucleolus

17. Centrioles are associated with?

    a) DNA synthesic   b) Reproduction

    c) Spindle formation   d) Respiration

18. Main difference between animal cell and plant cell is due to?

    a) chromosome   b) Ribosome

    c) Lysosome   d) none of these

19. Animal cell lacking nuclei would also lacking?

    a) Chromosome   b) Ribosome   c) Lysosome   d) none of these

20. Plasmolysis occurs due to?

    a) Absorption   b) Endosmosis

    c) osmosis   d) Exosmosis

21. A plant cell becomes turgid due to?

    a) Plasmolysis   b) Exdosmasis

    c) Endosmosis   d) Electrolysis

22. Solute Concentration is higher in the external solution called?

    a) Hypotonic   b) Isotonic

    c) Hypertonic   d) None of these

23. A cell placed in Hypertonic solution will?

    a) Swell up   b) Shrink

    c) No Change in shape, size   d) show plasmolysis

24. The radiant energy of sunlight is converted to chemical energy and is stored as?

    a) AMP   b) ADP

    c) ATP   d) APP

25. Which of the following organelle does not have membrane?

a) Ribosome   b) Nucleus

c) Chloroplast   d) none of these

26. Root hair absorbs water from soil through?

a) Osmosis   b) Active transport

c) Diffusion   d) Endocytosis

27. The number of lenses in compound light microscope is?

a) 4   b) 3   c) 1   d) 2

28. Which cell organelle is not bounded by a membrane?

a) Ribosome   b) Lysosome   c) ER   d) Nucleus

29. Which of the following cellular part possess a double membrane?

a) Nucleus   b) Chloroplast   c) Mitochondrion   d) All of the above

30. Cristae and Oxysomes are associated with?

a) Mitochondria   b) Plastids

c) Plasma Membrane   d) Golgi Apparatus

31. Karyotheca is another name of?

a) Nuclear envelope   b) Nucleus   c) Nuclear pores   d) Nucleolus

32. Cell organelle that acts as supporting skeletal framework of the cell is?

a) Golgi apparatus   b) Nucleus   c) Mitochondria   d) ER

33. Plastids are present in?

a) Animal cell only   b) Plant cell only

c) Both   d) None of these

34. Cell wall of plant is chiefly composed of?

a) Hemi cellulose   b) Cellulose c) Proteins   d) none of these

35. Intercellular connections of plant cells are called

    a) Middle lamella   b) Micro fibrils   c) Matrix   d) Plasmodesmata

36. Genes are located on the?

    a) Chromosomes   b) Nucleoulus

    c) Nuclear membrane   d) Plasma membrane

37. Chromatin Consists of?

    a) RNA   b) DNA   c) RNA& Histones   d) DNA & Histones

38. Name of the process that requires energy provided by ATP?

    a) Diffusion   b) Osmosis c) Active Transport   d) Plasmolysis

39. Vacuoles are present in?

    a) Animal cells   b) Plant Cells   c) Both   d) None of these

40. The cell is the fundamental structural unit of?

    a) Non living organisms   b) Living organisms

    c) Both   d) None of these

41. Who saw that the cork resembled the structure of a honeycomb consisting
    of many little compartments?

    a) Robert Hooke   b) N.S. Bose c) M.K. Chandran   d) None of these

42. ------------- is a substance which comes from the bark of a tree?

    a) Cork   b) torck   c) Metal   d) None of these

43. Cell is a latin word for?

    a) a little room   b) a big room

    c) a little basement   d) None of these

44. Hook made ------------ chance observation through a self designed
    microscope?

    a) Slice of cork   b) Small of cork   c) None of these

45. The very first time that someone had observed that living things apper to consist of separate units?

a) Hooke   b) Einstein   c) Golgi   d) none of these

46. Cell was first discovered by Robert Hooke in?

a) 1666   b) 1665   c) 1667   d) 1668

47. Who discovered the nucleus in the cell?

a) Robert Hook   b) Leeuwen Hook   c) Robert Brown   d) None of these

48. Purkinje in ----------- coined the term protoplasm for the fluid substance of the cell?

a) 1837   b) 1836   c) 1839   d) None of these

49. Which theory tells that all the plants and animals are composed of cells and that the cell is the basic unit of life?

a) Cell theory   b) Plant theory

c) Animal theory   d) None of these

50. Group of cells having similar structure and function are termed as?

a) Tissue   b) Organ System   c) Organ   d) Living organism

Answers:

QUES.	ANS.	QUES.	ANS.	QUES.	ANS.	QUES.	ANS.	QUES.	ANS.
1. (a)	2.(d)	3. (d)	4. (b)	5. (d)	6. (d)	7. (b)	8. (a)	9. (c)	10. (c)
11. (d)	12. (b)	13.(c)	14. (d)	15. (a)	16. (c)	17. (c)	18. (a)	19. (a)	20. (d)
21. (c)	22. (c)	23. (a)	24. (c)	25. (a)	26. (a)	27. (d)	28. (a)	29. (d)	30. (a)
31. (a)	32. (d)	33. (b)	34. (a)	35. (d)	36. (a)	37. (d)	38. (c)	39 (b)	40. (b)
41. (a)	42. (a)	43. (a)	44. (a)	45. (a)	46. (b)	47. (c)	48. (c)	49. (a)	50. (a)

2. TISSUE

SOME IMPORTANT POINTS

- ➤ Tissues are made up of cells.
- ➤ Millions of cells carried out a particular function in the body hence this group of cells is known as tissue.
- ➤ Mainly plant tissues are of two types :
1. Meristmetic tissue
2. Permanent tissue
- ➤ Merismetic tissue is divided into three parts.
1. Apical meristem
2. Inter calary meristem
3. Lateral meristem
- ➤ Apical meristem is found at the growing tips of roots and shoot to increase the length of stem and root.
- ➤ Permanent tissue are divided into two types :
1. Simple permanent tissue
2. Complex permanent tissue
- ➤ Parenchyma, Collenchyma, choleranchyma and sclerenchyma are the types of simple permanent tissue.
- ➤ Xylem and phloem are the type of complex permanent tissue.
- ➤ Xylem transports water and minerals from the root to the whole body and phloem transports food from leaves to the whole body of the plant.
- ➤ Animal tissues are mainly of four types :
1. Epithelial tissue
2. Connective tissue
3. Muscular tissue
4. Nervous tissue
- ➤ According to the structure epithelial tissue are divided as squamous, cuboidal, columnar, ciliated, and glandular.
- ➤ Different types of connective tissue are found in our body includes areolar tissue, adipose tissue, bone, tendon, cartilage and blood.

- Nervous is the functional unit of nervous tissue.
- Nervous tissue receives and conducts impulse. The impulses are read by brain.

2. Tissue

1. The unicellular organism has?

   (a)Single cell (b) Multiple cell (c) Both (a) & (b) (d) None of these

2. The multicellular organism is mainly?

   (a)Man   (b) Yeast   (c) Bacteria   (d) None of these

3. The activities perform by single called organism are?

   (a)Digestion   (b) Movements   (c) Both (a) & (b) (d) None of these

4. A tissue may be defined as a?

   (a)Group of cells (b)Single cells (c)Both(a)&(b)   (d)None of these

5. The cell walls of meristematic tissue are made up of?

   (a)Cellulose   (b) Protien   (c)Fats   (d) None of these

6. The lateral meristem is located at the growing?

   (a)Apices   (b)Leaf   (c)Stem   (d) None of these

7. The merestimatic tissue are responsible for?

   (a)Linear growth of an organ   (b) Increase in diameter

    (c)Growth of leaves   (d) None of these

8. The intercalary meristem responsible for?

   (a)Growth of leaves (b)Growth of leaves (c)Both(a)&(b)   (d) None of these

9. The living permanent cells are large and posses?

   (a)Thin walls   (b)Thick walls   (c)Both(a)&(b)   (d) None of these

10. Permanent tissue may be at?

    (a)Fixed position   (b)Dividing continuously (c)Both (a)&(b) (d) None of these

11. Food and water are conducted in plants through?

(a)Vascular tissue   (b)Mascular tissue   (c)Both(a)&(b)   (d) None of these

12. The supported tissue are mainly present in?

(a)Animals   (b)Plants   (C)Both(a)&(b) (d) None of these

13. Differentiation in the process of taking up a permanent?

(a)Shape   (b)size   (c)Function   (d) None of these

14. Simple permanent tissue are composed of?

(a)Only one type of cell (b)Variety of cell (c)Mixed cell (d) None of these

15. The flexibility in plants is due to the?

(a)Collenchyma (b)Parenchyma (c)Sclerenchyma   (d) None of these

16. In aquatic plants byouancy by which plants can float is provided by?

(a)parenchyma (b)Sclerenchyma (c)Collenchyma (d) None of these

17. To perform photosynthesis chlorophyll is contain in?

(a)Chlorenchyma (b)Collenchyma (c)Sclernchyma (d) None of these

18. Easy bending in different parts of plants without breaking is due to?

(a)Collenchyma (b)Parenchyma (c)Chlorenchyma (d) None of these

19. Collenchyma provided some functions to the organs where it is found these

Functions are?

(a)Flexibility (b)Elasticity (c)Both(a)&(b) (d) None of these

20. Sclerenchyma tissue help the plant to make it?

(a)Hard   (b)Soft   (c)Sniff   (d) None of these

21. The husk of a coconut is made up of?

    (a) Parenchyma   (b) Chlorenchyma   (c) Collenchyma   (d)Sclerenchyma

22. The walls of sclerenchyma tissue are thickened due to:
    (a)Proteins   (b)Lignin   (c)Chloroplast   (d)None of these.

23. Sclerenclhyma tissue is present in:

    (a)Stems   (b)The covering of seeds,nuts

    (c)Veins of leaves   (d) All of these

24. A little or no number of protoplasm is present in:

    (a)Parenchyma tissue   (b)Scelerenchyma tissue

    (c)Chlorenchyma tissue   (d) None of these

25. The outermost cell layer of the cell is ?

    (a)Epidermis   (b)Cytoplasm   (c)Stomata   (d) None of these

26. Cells of epidermis on the ariel parts helps again?

    (a)Loss of water (b)Cytoplasm (c)Both(a)&(b)   (d) None of these

27. Small pores in the epidermis of leaf are?

    (a)Cytoplasm   (b)Stomata   (c)Guard cell   (d) None of these

28. Each stomata is guarded with two kidney shaped cell called?

    (a)Epidermis cells (b)Plants cell (c)Guard cell (d) None of these

29. The opening and closing of stomata is regulated by?

    (a)Epidermal cell   (b)Plant cell (c)Guard cell   (d) None of these

30. The cell of cork of a tree are?

(a)Dead   (b)A live   (c)Small   (d) None of these

31. As plants grow older epidermis of stems are replaced by?

(a)Secondary meristem (b)Cytoplasm (c)Cork (d) None of these

32. The chemical that makes the cork wall impervious to gases and water is?

(a)Suberin   (b)Ammonium   (c)Both(a)&(b)   (d) None of these

33. Complex tissue are composed of?

(a)One type of cell   (b) More than one type of cell

(c) Mixed type of cell   (d) Both(b)&(c)

34. The minerals and water from root to different parts of plants is conducted by?

(a)Xylem   (b)Pholem   (c)Branches   (d) None of these

35. The food material from the leaves to different parts of plants is conduct by?

(a)Xylem   (b)Pholem   (c)Roots   (d) None of these

36. The parts of xylem in a plant are?

(a)Tracheids   (b)Vessels   (c)Xylem ,Parenchyma   (d)All of these

37. The food stored in the form of starch fat in ?

(a)Xylem Parenchyma   (b) Vessels   (c) Tracheids   (d) None of these

38. The constitute of phloem are?

(a)Sieve tube   (b)Companion   (c)Vessels   (d)Both(a)&(b)

39. Xylem consist both living and non-living?

(a)Beings   (b)Cells   (c)Chloroplast   (d) None of these

40. In phloem food material from the leaves carry to storage organ and then

growing region of plant body by?

(a)Sieve tube   (b)Pollen tube   (c)Conducting tube   (d) None of these

41.   The passage of materials is controlled by?

(a)Conducting cell   (b)Companion cell   (c)Epidermal cell   (d) None of these

42.   The tissue found in our body ?

(a)Blood   (b)Xylem   (c)Muscles   (d) Both(a)&(b)

43.   Blood is a which type of tissue?

(a)Epithelial tissue   (b)Connective tissue   (c)Muscular tissue   (d) None of these

44.   The covering or protecting tissue in the animal body are?

(a)Epithelial tissue   (b)Connective tissue   (c)Nervous tissue   (d) None of these

45.   The skin which protects the body is made up of?

(a)Squamous epethilial   (b)Cuboidal epithelial

(c)Columnar epethilial   (d) None of these

46.   In the respiratory tract the tissue that have cilia with pair like projection is?

(a)Squamous epithelial   (b)Stratified squamous

(c)Columnar epithelial   (d) None of these

47.   The tissue that makes the lings of kidney tubules ,ducts,of salivary glands is?

(a)squamous epithelium   (b)Cuboidal tissue

(c)Columnar epithelium   (d)) None of these

48.   Bone a hard and non flexible tissue is an example of?

(a)Columnar tissue   b)Connecting tissue

(c)Nervous tissue   (d)) None of these

49. Blood has been a fluid matrix called ?

   (a)Plsma   (b)RBC   (c)WBC   (d)Protiens

50. Two bones can be connected to each other by connective tissue called?

   (a)Ligament   (b)Tendons (c)Muscles (d)) None of these

51. The tissue connects muscles to bones ?

   (a)Ligament   (b)Tendon (c)Muscles   (d)) None of these

52. The tough ,smooth and flexible connective tissue hat present in nose ,ears

   trachea is ?

   (a)Ligament   (b)Cartridge   (c)Tendons (d)) None of these

53. The tissue that found between the skin and muscles in the bone marrow ,

   Supports internal organs is ?

   (a)Ligament   (b)Cartridge   (c)Areolar   (d)) None of these

54. The tissue that found below the skin and store fats of our body is?

   (a)Ligament   (b)Adipose (c)Tendons   (d)) None of these

55. Which tissue cause movement in our body?

   (a)Supporting tissue   (b)Mascular tissue

   (c)Nervous tissue   (d)) None of these

56. Example of voluntary action muscle is?

   (a)Muscles in our body   (b)Contratiction of blood vessel

   (c)Sneezing   (d)) None of these

57. The example of involuntary action muscle is?

   (a)Muscles in our limbs (b)The iris of eye (c)Lymph (d)Both (a)&(b)

58. The involuntary muscles of heart is called?

(a)Lymph   (b)Cardiac

(c)Neuron muscles   (d)) None of these

59. The compositons of tissue are?

(a)Brain   (b)Spinal cord   (c)Nerves   (d)All of these

60. The cells of Nervous tissue is called?

(a)Neurons   (b)Nerve cells

(c)Both (a)&(b)   (d)) None of these

61. The longest cell of our body is?

(a)Neuron   (b)Lymph

(c)Forelimp   (d)) None of these

62. A Nerve cell consists of different parts?

(a)Cell body   (b)Axon

(c)Dendrites   (d)All of these

Answers :

Q	A	Q	A	Q	A	Q	A	Q	A	Q	A	Q	A	Q	A	Q	A
1	A	8	A	15	A	22	B	29	C	36	D	43	B	50	A	57	D
2	A	9	A	16	A	23	D	30	A	37	A	44	A	51	B	58	B
3	A	10	A	17	A	24	B	31	A	38	D	45	A	52	B	59	D
4	A	11	A	18	A	25	A	32	A	39	B	46	C	53	C	60	C
5	A	12	B	19	C	26	C	33	B	40	A	47	B	54	B	61	A
6	C	13	D	20	D	27	B	34	A	41	B	48	B	55	B	62	D
7	A	14	A	21	D	28	B	35	B	42	D	49	A	56	A		

3. DIVERSITY IN LVING ORGANISMS

SOME IMPORTANT POINTS

➤ Classification helps us to read the diversity of life forms easily.

➤ The major characteristics taken for classify organisms are :

    1. Whether they are made up of prokaryotic or eukaryotic cells.

    2. They are made up of single or multi cells.

    3. Whether they made their food by own or not.

➤ The main credit of classification of organisms goes to Robert Whittaker.

➤ The organisms are classified into 5 kingdoms.

    1.Monera

    2.Protista

    3.Fungi

    4.Plantae

    5. Animalia

➤ The organisms are classified into five kingdoms:

    1. Thallophyta

    2. Bryophyta

    3. Pteridophyta

    4. Gymnosperm

    5. Angiosperm

➤ The kingdom anamilia are divided into 10 groups:

    1. Porifera

    2. Coelentreta

    3. Platyhelminthes

4. Nematode

5. Annelida

6. Arthopoda

7. Mollusca

8. Echinodermata

9. Protochordata

10. Vertebreta

➤ Vertebrata are divided into five sub classes:

1. Pisces

2. Amphibia

3. Reptelia

4. Aves

5. Mamallia

➤ Carolus linnaeus introduced the system of scientific nomenclature in the 18th century.

3. Diversity in living organisms

1.   Who wrote the 'THE ORIGIN OF SPECIES'?

   (a)Robert hooke   (b)Robert frost

   (c)Robert Charles Darwin   (d)Robert brown

2.   Which we called the region of mega diversity?

   (a) Tropic of cancer   (b) Tropic of capricon

   (c)Both(a)&(b)   (d)None of these

3.   Who classify the organism into 5 kingdoms ?

   (a)Robert whittaker   (b)Carl woese

   (c)Frust haekel (d)Robert   (d) darwin

4.   According to the whittaker the 4th kingdom is?

   (a)Animalia   (b)Plantae   (c)Fungi   (d)Protista

5.   Mycoplasma is an example of which kingdom ?

   (a)Protista   (b)Monera   (c)Animalia   (d)Fungi

6.   Cyarobacteria is known as ?

   (a)Mycoplasma   (b)Euglera

   (c)Blue-green algae   (d)Ipomea

7.   Diatoms is an example of which kingdom ?

   (a)Monera   (b)Fungi

   (c)Plantae   (d)Protista

8.   Fungi have cell walls made of tough complex sugar called ?

   (a)Chitin   (b)Cytosome

(c)Crystals   (d)Cycas

9.   Plants belongs to which group called algae ?

        (a)Bryophyta   (b)Petridophyta

        (c)Thallophyta   (d)None of these

10.   Which group is called the amphibians of the plant kingdoms ?

        (a)Thallophyta   (b)Bryophyta   (c)Petridophyta   (d)None of these

11.   Which of the following organisms cannot have cell wall ?

        (a)Neem (b)Spirogyra (c)Deodar (d)Hydra

12.   The animals have holes or pores all our the body are known ?

        (a)Coelantrata   (b)Phatyhelmenthes   (c)Arthopoda   (d)Porifera

13.   In porifera which helps in circulating water through out the body ?

        (a)Chitn   (b)Holes   (c)Pores   (d)(b)&(c)are same

14.   Spongilla are mainly found in ?

        (a)Terestrial habitat   (b)Marine habitat

        (c)Both (a)&(b)   (d)None of these

15.   Which of the following animal body made up of two ?

        (a)Porifera   (b)Sycon   (c)Planarians   (d)Hydra

16.   Which of these animal is covered with the a hard outside skeleton ?

        (a)Sycon   (b)spongilla (c)Hydra   (d)both (a) & (b)

17.   Which of the these group of animalia a pseudocoeln in present ?

        (a)Porifera   (b)Nematoda

        (c)Platyhelminthesis   (d)Coelentreta

18. Leeches and earthworms are example of ?

    (a)Arthopoda   (b)Annelida   (c)Nematoda   (d)Coelentreta

19. The largest group of animals ?

    (a)Arthopoda   (b)Nematoda   (c)Annelida   (d)None of these

20. Which of these have jointed legs?

    (a)Musca   (b) Butterfly   (c) None of these   (d)Both (a) & (b)

21. Palaemon Arena and musca are the example of?

    (a)Annelida   (b)Nematoda   (c)Arthopoda   (d)None of these

22. Pariplanta is the scientific name of?

    (a)Housefly   (b)Butterfly   (c)Musca   (d)Cockroach

23. Aranea is the scientific name of?

    (a)Scorpion   (b)Spider   (c)Musca   (d)Cockroach

24. Asterias is the scientific name of?

    (a)Cockroach   (b)Butterfly   (c)Sea wichins   (d)Starfish

25. Starfish and sea urchins are the example of?

    (a) Arthopoda   (b)Echinodermata   (c)Mollusca   (d)Nematoda

26. Which of the following is not the features of vertebrate?

    (a) Have a notochord   (b) Have a dorsal nerve cell

    (c) Have a gill bouched   (d) Are tripoblastic

27. Vertebrate are divides into _____ classes

    (a) 6   (b)5   (c)7   (d)4

28. Which of the following have only two chambered heart ?

(a)Petrio volitans   (b)Hyle   (c)Rana tigriana   (d)All of above

29.   Hyla belongs to which class ?

(a)Pisces   (b)Reptilia   (c)Amphibia   (d)All of above

30.   Scientific name of frog is?

(a)Hyla   (b)Rana   (c)Salamander   (d)Rana tigriana

31.   Amphibians are breath through?

(a)Lungs   (b) Gills   (c) Liver   (d) Both (a) & (b)

32.   Crocodile have _____chambers heart

(a)3   (b)4   (c)2   (d)None of these

33.   Crocodiles are belong to which group ?

(a)Reptilia   (b) vertebrate   (c)Arthopoda   (d)Plantae

34.   Which of these have mammary gland ?

(a)Whale   (b) Bat   (c) Human   (d)All of above

35.   Moss and Marelantra are the example of?

(a)Thallophyta   (b)Bryophyta

(c)Petridophyta   (d)Phanerograms

36.   Which of these have specialised tissue for conduction of food and water?

(a)Funaria   (b)Riccia   (c)Fern   (d)Chara

37.   Ferns ,Marsila and horse tails are the example of ?

(a)Arthopoda   (b)Mollusca   (c)Petridophyta   (d)Bryophyta

38.   Pires and deodar are the example of?

(a)Angiosperm   (b)Phanerogams

(c)Teridophyta   (d)Thallophyta

39.   Hemidactylus is the scientific name of?

(a)Sparrow   (b)Lizard   (c)Pigeon   (d)Housefly

40.   The scientific name of Flying lizard is ?

(a)Peluin   (b)Draco   (c)Both (a) & (b)   (d)Hyla

41.   Which of the following is heterotrobh ?

(a)Musca (b)Ipomea (c)Moss (d)Ulothrix

42.   Anabaena is the example of?

(a)Animalia   (b)Plantae   (c)Protista   (d)Monera

43.   Scientific name of human is?

(a)Homo   (b) sapens   (c) Homospiens   (d) All of above

44.   The system of scientific name is introduce?

(a)Charles Darwin   (b) Carolous linneus

(c) Both (a) & (b)   (d) None of these

45.   The scientific name of white stork is ?

(a) Ciconia   (b)Ciconia-Ciconia

(c) None of these   (d) Camel

46.   The scientific name of dog fish is ?

(a)Scoliodon   (b)String ray   (c)Torpedo   (d)None of these

47.   Chameleon have _____ chamber heart
(a) 2   (b) 3   (c) 4   (d) None of these

48.   Male hippocampus is the scientific name of?

(a) Flying fish   (b) Sea horse   (c) String ray   (d) Dogfish

49. Blanoglassus is the example of?

(a)Nematoda   (b)Arthopoda   (c) Protochordata   (d) Annelida

50. Herdmania and Amphioxus is the example of?

(a) Nematode   (b) Arthropod   (c)Protochordata   (d) Annelida

51. Liver fluke belongs to which group?

(a)Nematoda   (b) Platyhelmenthes

(c) Annelida   (d) Arthopoda

ANSWERS:

Q	A	Q	A	Q	A
1	C	18	B	35	B
2	C	19	A	36	C
3	A	20	D	37	C
4	B	21	C	38	B
5	B	22	D	39	B
6	C	23	B	40	B
7	D	24	D	41	A
8	A	25	B	42	D
9	C	26	C	43	C
10	B	27	B	44	B
11	D	28	A	45	B
12	D	29	C	46	A
13	D	30	D	47	B
14	B	31	D	48	B
15	D	32	B	49	C
16	D	33	B	50	C
17	B	34	D	51	B

4. WHY DO WE FALL ILL

SOME IMPORTANT POINTS

- Health is a state of physical, mental, and social well being.
- The health of an individual is dependent on his/her physical surrounding and economic status.
- Diseases are characterised as a acute and chronic disease on their duration.
- Disease Kala-azar is caused by Leis mania.
- Viral disease is transmitted by contact by form a sick person to healthy person.
- Infectious agents are spread through air, water or vectors.
- Prevention of disease is best its successful treatment.
- Infectious disease can also be prevented by using immunisation.
- The category to which a disease-causing organism belongs decides the types of treatment.

4. WHY DO WE FALL ILL

1.   The musculoskeletal system which is made up of.

       (a) Bones   (b) muscles   (c) bone and muscles   (d) none of these

2.   Musculoskeletal system holds the body part together and helps the body

       (a)move   (b)stop   (c)rest   (d)none of these

3.   A state of being well enough to function well physically.

       (a)health   (b)disease   (c)none of these

4.   Kidney is filtering.

       (a) beating   (b)fluring   (c)wine   (d)thinking

5.   Energy and raw material are needed from ---------------- the body.

        (a)inside   (b)outside   (c)none of these

6.   What is a necessity for cell and tissue functions?

        (a) Sun light   (b) air   (c) food   (d) oxidation

7.   The environment includes the.

       (a)chemical environment   (b)physical environment

       (c)biological environment   (d)none of these

8.   Health is at risk in a cyclone in -------------- ways.

       (a) short   (b)small   (c)many   (d)big

9.   Our physical environment is decided by our.

       (a) Physical environment

       (b) Chemical environment

(c) Social environment

(d) biological environment

10.   Food will have to be earned by doing.

(a) fighting   (b)energy   (c)work

11.   Some diseases last for only very short periods of time called .

(a)chteronic diseases   (b)acute disease

(c)cuteronic diseases   (d)none of these.

12.   Some diseases last for a long time even as much as a lifetime called.

(a)chronic disesases   (b)acute diseases

(c)cuteronic diseases   (d)none of these

13.   Which is the example of chronic disease?

(a)Tuberculosis   (b)fever   (c)Aids   (d)a and c

14.   Which is the example of chronic disease.

(a)fever   (b)T.B   (c) all diseases   (d)none of these

15.   Loose motions is an example of

(a) acute diseases   (b)chronic diseases

(c) cuteronic diseases   (d)none of these

16.   The disease that is non communicable is.

(a) Malaria   (b) marasmus   (c) AIDS   (d) jaundice

17.   Malaria is caused by a?

(a)protozoan   (b)fungi   (c) virus   (d) bacteria

18. The hungi of the isabgol seed with water produces relief from?

    (a)malaria   (b) flue   (c)diarrhea

19. Oral rehydration solution does not contain?

    (a) sodium   (b)sodium bicarbonate

    (c) Sodium cyanide   (d)glucose

20. The vitamin that is not fat soluble is ?

    (a)vitamin A   (b)vitamin B complex

    (c)vitamin D   (d)vitamin E

21. Xerophthalmia is caused due to the deficiency of?

    (a)vitamin A   (b)vitamin B   (c)vitaminE

22. The 4D-syndrome characterizes the following disease?

    (a)pellagra   (b)scurvy   (c)beriberi (d)xeropthalima

23. Maize interferes with the absorption of?

    (a)ascorbic   (b)nicpotinic acid

     (c)thiamine   (d)iodine

24. Sun light enhances the production of?

    (a)vitamine A   (b)vitamine B

    (c)vitamine C   (d)vitamine D

25. The proposed two in one salt condium iodine and

    (a)sodium   (b)potassium   (c)iron   (d)manganese

26. An insect which transmits a disease is known as?

    (a)intermediate host   (b)parasite   (c)vector   (d)prey

27. A chronic case of a diseases denotes?

(a) Severe attack of a disease

(b) Mild attack of the diseases

(c) Disease occurs for a very long period

(d) all of these

28. Which one of the diseases is not communicable?

(a)Typhoid   (b)diabetes   (c)hypertension   (d)helminthes

29. Which one of the diseases is none communicable?

(a)typhoid   (b)leprosy   (c)measles   (d)leukemia

30. Congenital diseases are those which?

(a) Are deficiency diseases

(b) Are present from time of birth

(c) Are spread from man to man

(d) Occur during life time

31. BCG vaccine is used to curb?

(a)Pneumonia   (b) Tuberculosis   (c)Polio   (d) Amoebiasis

32. AIDS virus has?

(a)Single strand DNA   (b) double strand DNA

(c)Single strand RNA   (d) double strand RNA

33. Causative agent of T.B.is?

(a)defective liver   (b)defective thymus

(c)AIDS virus   (d) weak immune system

34. Immuno deficiency syndrome could develop due to ?

(a) Defective liver   (b) defective thymus

(c)AIDS virus   (d)weak immune system

35. T.B. is cured by?

(a) grise ofulvin   (b) ubiquinone   (c) encitol   (d) streptomycin

36. AIDS is due to?

(a) Reduction in number of helper t- cell

(b) Reduction in number killer   t-cell

(c) Auto immunity

(d) non production of intererons

37. Typhoid is caused by?

(a)Escherichia   (b)giardia   (c)salmonella   (d)shigella

38. Which of the following is a mismatch?

(a) Leprosy – bacterial infection   (b) AIDS–bacterial infection

(c) Malaria – protozoan infection   (d) elephantiasis- nematode infection

39. Calcium deficiency occurs on the absence of vitamin?

(a) D   (b)A   (c)C   (d)B

40. Fever, delirium, slow pulse, abdominal tenderness and rose colored rash

Indicate the disease?

(a)typhoid   (b) measless   (c)tetanus   (d)chicken pox

Answers:

QUES.	ANS.	QUES.	ANS.	QUSE.	ANS.	QUES.	ANS.
1	C	11	B	21	A	31	B
2	A	12	A	22	A	32	C
3	A	13	D	23	B	33	B
4	C	14	A	24	D	34	C
5	A	15	A	25	C	35	D
6	C	16	B	26	C	36	A
7	B	17	A	27	C	37	C
8	C	18	C	28	A	38	B
9	C	19	C	29	D	39	A
10	C	10	B	30	b	40	A

5. LIFE PROCESS

SOME IMPORTANT POINTS

- All living things perform basic life process like, growth, digestion, excretion, respiration, circulation, etc.
- The basic functions perform by living organisms for their survival and body maintenance are called life process.
- There are mainly two models nutrition :
1. Autotrophic   2.Heterotrophic
1. Heterotrophic: They synthesis their food by own by the process of photosynthesis.
2. Raw materials required for photosynthesis is Co_2 and H_2O.
3. Tiny pores present on the surface of the leaves are known as stomata.
4. Stomata performs a special function in the exchange of gases and loses and large amount of water during transpiration.
5. Human alimentary canal originates from the mouth and to the anus.
6. Respiration involves exchange of gases and break down of simple food in order to release energy.
7. Breakdown of glucose by various path ways:
- Respiration in plants is simpler than the respiration in animals. Gaseous exchange occurs through.
1. Stomata in leaves
2. Lenticels in stem
3. General surface of the roots
- The process of transport food and oxygen, etc. Supply in the multicellular organisms is known as transportation.
- Human circulatory system include :
1. Heart
2. Arteries & Veins
3. Blood & Lymph
- In human beings blood travels twice through the heart to the body
1. Pulmonary Circulation
2. Systemic Circulation

- Blood contains plasma, R.B.C, W.B.C and platelets.
- Transportation in plants takes place by xylem and phloem tissue.
- Transport of food from leaves to the rest of the body is known as translocation.
- The process of removing harmful substances from the body is known as excretion.
- Excretory system includes a pair of kidney, a pair of ureter, a urinary bladder and a urethra.
- Nephron is the filtration unit of kidney.
- The process of purify blood by an artificial kidney is known as Haemodialysis.

5. LIFE PROCESS

1. Name the organic compounds used by plant to make their food?

   (a)Water and minerals   (b)Water and CO_2

   (c)Water and Minerals   (d) none of these

2. Plant converted carbon dioxide and water into in the presence of

   Sunlight & chlorophyll?

   (a)Proteins   (b) Carbon monoxide

   (c)Vitamins   (d) Carbohydrates

3. Carbohydrates are stored in plant body in the form of?

(a)Sugar   (b)Starch   (c)Fats   (d)Glycogen

4.   The food we eat is stored in our body in the form ?

(a)Starch   (b) Glycogen   (c) Fats   (d) Sugar

5.   The molecular formula of glucose is?

(a)$C_6H_{12}O_6$ (b) $C_6H_2O_2$ (c)$C_6H_{12}O_6$ (d)$C_6H_3O_5$

6.   The molecular formula of sugar is?

(a) $C_6H_{12}O_6$ (b) $C_{12}H_{22}O_{11}$ (c) $C_{12}H_{21}O_9$ (d)$C_9H_{22}O_{11}$

7.   Which event is not occurring during Photosynthesis?

(a) Aborsoption of chlorophyll by light energy.

(b) Spiliting of water molecules .

(c) Conversion of light energy into chemical enrgy.

(d) Reduction of carbon dioxide into carbohydrates.

8.   Plants take which gas for respiration?

(a)CO_2 (b)O_2 (c)CO   (d)MgO

9.   _____ contains chlorophyll?

(a)Xylem   (b) Chloroplast   (c) Phloem   (d) Both (a) & (b)

10.   The tiny pores present on the surface of leaves are known?

(a)Poral   (b) Stomata   (c) Stomach   (d)Both (b)&(c)

11.   The opening and closing of the pore is a function of?

(a)Stomata   (b) Guard cell(c)Stomatal pore (d)None of these

12.   For the synthesis of protein and other compound which metal is

essential?

(a)NO_2 (b) N (c)CO_2 (d)O_2

13. Which bacteria are useful for nitrogen fixation?

(a)Rizopus (b)Rizoed (c)Rizobium (d)Rizobesilus

14. The alimentary canal is about?

(a)6m (b)9m (c)9Km (d)8Km

15. The small intestine is about?

(a)6m (b)6.5m (c)6.3 (d)6.6

16. Saliva is secreted by?

(a)Teeth (b) Salivary gland (c) Tongue (d)Both(a)&(b)

17. Saliva makes the _____

(a)Passage rough (b) Passage smooth

(c)Passage wet (d) Both(b)&(c)

18. The movement that push the food forward to the oesophagus?

(a)Peristaltic movement (b)Peritaltic movement

(c)Periataltic movement (d)None of these

19. Mucus protect_____ by the action of HCl?

(a)Inner lining of the stomach (b)Outer lining of the stomach

(c)All of the body (d)Both (b)&(c)

20. Which gland release pepsin?

(a)Bile duct (b) Gastric gland

(c)Gustatory gland (d) Pancreatic gland

21. The exit of food from the stomach is regulated by?

(a)Contract muscles   (b) Peristaltic muscles   (c)Sphincter muscles   (d)Stomachal muscles

22. Which is the longest part of the alimentary canal?

(a)Large intestine   (b) Small intestine

(c)Stomach   (d) Anus

23. Which is the site of the complete digestion of carbohydrates,fats

and proteins?

(a)Stomach   (b) Small intestine   (c) Large intestine   (d) Anus

24. The enzyme pepsin works well in?

(a)Alkaline medium   (b)Acidic medium

(c)Neutral medium   (d)Basic medium

25. The pancreatic enzyme work well in_____?

(a)Acidic medium   (b)Acid medium

(c)Both(b)&(c)   (d)Alkaline medium

26. Bile juice is _____ in nature?

(a)Acidic   (b) Both(b)&(c)

(c)Alkaline   (d) Neutral

27. Enzyme like trypsin and lipase is secreted by?

(a)Intestinal juice   (b) Pancreas

(c)Gastric gland   (d)Bile duct

28. The inner lining of the small intestine has numerous fingers like

Projection called?

(a)Valli   (b)Villi

(c)Mount   (d)Velli by anus

29.   The exit of the waste material is regulated by ?

(a)Anus   (b) Anal sphincter

(c)Anal   (d) Anus sphincter

30.   Dental caries or tooth decay causes softening of ?

(a)Dentine   (b) Enamel

(c)Tooth   (d) Both (a)&(b)

31.   Saliva is _____ is nature

(a)Acidic   (b) Neutral

(c)Basic   (d) Both (b)&(c)

32.   The breakdown of glucose into pyruvate takes place in ?

(a)Cell   (b)Stomael

(c)Tissue   (d)Cytoplasm

33.   The breakdown of pyruvate into co_2 ,water and energy takes place in ?

(a)Yeast   (b)Muscle

(c)Lysosome   (d) Mitochondria

34.   _____ is the energy currency of most cellular respiration?

(a)ADP   (b)ATP   (c)APP   (d)Both (a) &(b)

35.   Which provide a surface area for the exchange of gasses?

(a)Nostrils   (b) Alveoli   (c) Larynx   (d)Trachea

36.   The respiratory pigment present in the human body is?

(a)Haemoglobin (b)Tryspin (c)Chloroplast (d)Cytokinnin

37.   The blood pressure is measured by?

(a)Sphygometer   (b)Shygomomanometer

(c)Sphygorameter   (d)Bloodpremeter

38.   The SI unit of blood pressure is?

(a)N   (b)Pa   (c)Pascal   (d)Both (b)&(c)

39.   Which can avoid the leakage of blood?

(a)Platelet   (b)Capillaries   (c)Aorte   (d)Both (b)&(c)

40.   The transport of water and animals from soil to plant body is done by?

(a)Lympta   (b) Lymph   (c)Xylem   (d)Phloem

41.   The transport of blood from leaves to other part of the plant is done by?

(a)Lymph   (b)Xylem   (c)Phloem   (d)Lympth

42.   Which of the following is not the part of the nephores?

(a)Bowman's capsule   (b)Glomerulus

(c) Collecting duct   (d) None of these

43.   Which of these following is not the part of digestive system ?

(a)Trachea   (b) stomach   (c) Food pipe   (d)None of these

44.   Which is the power house of cell?

(a)Lysosome   (b) ATP

(c)Mitochondria   (d) Golgi apparatus

45.   The kidney is the part of?

(a)Respiration   (b)Excretion   (c)Transportation (d)Nutrition

46.   The autotrophic mode of nutrition requires?

   (a)CO_2 (b) All of these   (c) Chlorophyll   (d)Sunlight

47.   The breakdown of the pryuvate to give ethanol takes place?

   (a)Muscles   (b) Mitochondria   (c)Lysosome   (d) Yeast

48.   The anus in human beings in a part of?

   (a)Transportation system (b)Alimentay canal

   (c)Respiratory system   (d) Excretion system

49.   By which process amoeba takes food from the outside?

   (a)Pseudopodia   (b)Endrecytosis   (c)Exocytosis   (d)Both (a)&(b)

Answers:

QUES.	ANS.	QUES.	ANS.	QUES.	ANS.	QUES.	ANS.	QUES.	ANS.

1	C	11	B	21	C	31	C	41	C
2	C	12	B	22	B	32	D	42	D
3	B	13	C	23	B	33	D	43	A
4	B	14	B	24	B	34	B	44	C
5	C	15	B	25	D	35	B	45	B
6	B	16	B	26	C	36	A	46	B
7	A	17	D	27	B	37	B	47	D
8	B	18	A	28	B	38	D	48	B
9	B	19	A	29	B	39	A	49	B
10	B	20	B	30	D	40	C	50	

6. CONTROL AND CO-ORDINATION

SOME IMPORTANT POINTS

➤ All information from our environment is detected by the specialised tips of nerve cells called receptors. These are usually our sense organ.

➤ The gap between two neuron is synapse.

➤ Reflex action is response without thinking or any conscious thought. This response is almost involuntary.

➤ The pathway involved in a reflex action is called a reflex arc.

➤ Spinal cord is a cylindrical structure and it about 45cm long.

➤ Human brain has 3 parts :

1. Fore brain: It is thinking part of brain.

2. Mid brain: Control reflexes involve eyes and ears.

3. Hind brain: Controls body equilibrium.

➢ Plant cell change shape changing the amount of water in them, resulting in swelling or shrinking.
➢ The movement towards light is phototrophic movement, towards water is hydrotropism, and toward gravity is geotropism.
➢ Chemotropism: It is the direction movement of plant part in response to chemical stimulus.
➢ Plant hormone auxin stimulates to growth and gibberellins promote stem elongation.
➢ Iodine is necessary for thyroid gland for the production of thyroxin hormone.
➢ Insulin regulates blood glucose level.

6. CONTROL AND CO-ORDINATION

1.   In animals control and coordination are provided by?

(a)Nervous system   (b) Muscular tissues

(c)Both (a) & (b)   (d) None of these

2.   The receptors which detects change in our environment located in our sense

Organ, these are?

(a)The nose   (b) The ear

(c) All of these   (d)The tongue

3.   The receptors which detect taste are?

(a)Olfactory receptors   (b) Gustatory receptors

(c)Both (a) & (b)   (d) None of these

4. The receptors which detect smells are?

(a)Olfactory receptors   (b) Gustatory receptors

(c) Both (a) & (b)   (d) None of these

5. The electrical impulse travels from to the?

(a)Axon   (b) Cell body

(c)Synapse   (d) None of these

6. The delivery of the impulses from neurons to other cells is allowed by?

(a)Axon   (b) Cell body

(c)Synapse   (d) None of these

7. Where information is acquired first in a neuron?

(a)Axon   (b) Dendrite

(c)Synapse   (d) None of these

8. The electrical impulse sets of the release at some chemical signals at?

(a)Axon   (b) Dendrite

(c)Synapse   (d) None of these

9. Nerves from all over the body meet in a bundle in the?

(a)Brain   (b) Neuron

(c)Spinal cord   (d) none of these

10. Reflex arc are generally formed in the?

(a)Brain   (b) Neuron

(c)Spinal cord   (d) none of these

11.   Which is a type of receptor of a body?

       (a)Muscle   (b) Skin

       (c) Foot   (d) none of these

12.   Which is a type of effectors of a body?

       (a)Muscle   (b) Skin

       (c) Foot   (d) none of these

13.   The central nervous system consists of?

       (a)Brain   (b) Spinal cord

       (c) Heart   (d) Both (a) & (b)

14.   Which is an example of voluntary action?

       (a)Sneezing   (b) Clapping

       (c)Coughing   (d) none of these

15.   Which is an example of involuntary action?

       (a)Sneezing   (b) Clapping

       (c)Talking   (d) none of these

16.   The communication between the central nervous system and other parts of s body is facilitated by?

       (a) Pripheral nervous system (b) Perial nrvous system

       (c) Perinial nervous system   (d) None of these

17.   The cranial nerves and the spinal nerves are consist of?

       (a)Pripheral nervous system (b)Perial nrvous system

(c)Capillary   (d)   None of these

18.   The sensory impulses from various receptors are received by?

        (a)Fore brain   (b) Mid brain

        (c)Hind brain   (d) None of these

19.   The main thinking part of the brain is?

        (a)Fore brain   (b) Hind brain

        (c)Mid brain   (d)None of these

20.   The decision hearing, movement of voluntary muscles, smelling, sight, and so

        on, are specified by ?

        (a)Fore brain   (b) Mid brain

        (c)Hind brain   (d) None of these

21.   A part of the hind brain is?

        (a)Cerebrum   (b) Cerebellum

        (c)Capillary   (d) Vertebral

22.   The involuntary action including blood pressure, salivation and vomiting are

        Controlled by?

        (a)Fore brain   (b)Mid brain

        (c)Hind brain   (d)None of these

23.   The part of brain which is responsible for voluntary actions?

        (a)Cerebrum   (b)Cerebellum

        (c)Capillary   (d)Medulla

24.   Which fluid acts as a shock absorbe in brain ?

(a)Andruistic fluid   (b)Paraphenocarpy

(c)Cerebrospinal fluid   (d) Adrenaline

25. The spinal cord protected by?

(a)Glliberllins   (b)Vertebral column

 (c)Parathyroid   (d) Gonads

26. Which is a sensitive plant?

(a)Mimosa's family   (b)Bryophyllum

 (c)Pisum Satvium   (d)None of these

27. Which part of sensitive plant involve in response to touch?

(a)Stem   (b) Root

(c)Leaves   (d)None of these

28. In phototropic movements plants respond towards?

(a)Gravity   (b)Light

(c)Air   (d)Heat

29. In geotropism movements plants respond towards?

(a)Gravity   (b)Light

(c)Air   (d)Heat

30. In hydrotropism movements plants respond towards?

(a)Gravity   (b) Light

(c)Water   (d) Heat

31. In chemotropism movements plants respond towards?

(a)Gravity   (b)Light

(c)Air   (d)Chemical

32.   Which hormone synthesized at the shoot tip of plant?

        (a) Auxin   (b) Cytokines

        (c)Gibberellins   (d)Abscisic acid

33.   Which harmone help in promoting growth in plants?

        (a)Auxin   (b) Cytokines

        (c)Gibberellins   (d)Abscisic acid

34.   Which harmone inhibits growth in plants?

        (a) Auxin   (b) Cytokines

        (c) Gibberelllins   (d) Abscisic acid

35.   Which hormone synthesized at the stem of plant?

        (a)Auxin   (b)Cytokinns

        (c)Gibberelllins   (d)Abscisic acid

36.   Which hormone causes increasing in heart beats?

        (a)Insulin   (b) Adrenaline

        (c)Thyroxin   (d)None of these

37.   The gland for which iodine is necessary?

        (a)Adrenal gland   (b) Pituitary gland

        (c)Thyroid gland   (d)None of these

38.   Which is a type of emergency gland?

        (a)Adrenal gland   (b)Pituitary gland

        (c)Thyroid gland   (d)None of these

39.   The hormone Which is responsible for growth in animals ?

(a)Insulin   (b) Thyroxin

(c)Adrenaline   (d) Pituitary

40.   One of the symptoms of goitre is ?

(a)Increassing fat   (b) Unexpected growth

(c) Swollen neck   (d)None of these

41.   The goitre disease is due to deficiency in ?

(a)Protien   (b)Iodine

(c)Carbohydrate   (d)Energy

42.   The growth hormones are secreted by ?

(a) Adernal gland   (b)Pituitary gland

(c)Thyroid gland   (d)Pancrease gland

43.   The hormone which regulating blood sugar levels is?

(a)Thyroxin   (b)Adernaline

(c)Insulin   (d)Estrogen

44.   The insulin hormone is secreted by?

(a)Adernal gland   (b)Pituitary gland

(c)Thyroxin gland   (d)Pancrease gland

45.   The hormone which is released which by testis gland?

(a)Oestrogen   (b)Testostrone

(c)Thyroxin   (d) Insulin

46.   Which harmone promote fruit ripening oin plants ?

(a) Auxins   (b) Gibberellins

(c) Cytokinns   (d) Ethylene

47.   The gap between two neurons is called a?

(a) Dendrite   (b) Synapse

(c) Impulse   (d) Axon

48.   Which gland is an endocrine glands ?

(a) Pancreas   (b) testis

(c) Both (a)&(b)   (d) None of these

49.   Which is not a plant?

(a) Auxin   (b) Oestrogen

(c) Cytokinns   (d) None of these

50.   The breathing rate increases due to increasing at?

(a) Diaphragm   (b) Rib muscles

(c) Both (a) & (b)   (d) None of these

Answers:

Q.	A.	Q.	A.	Q.	A.	Q.	A.	Q.	A.	Q.	A.	Q.	A.	Q.	A.	Q.	A.	Q.	A.
1	C	6	C	11	B	16	A	21	B	26	A	31	D	36	B	41	B	46	D
2	C	7	B	12	A	17	A	22	C	27	C	32	A	37	C	42	B	47	B
3	B	8	A	13	D	18	A	23	B	28	B	33	B	38	A	43	C	48	C
4	A	9	C	14	B	19	A	24	C	29	A	34	D	39	D	44	D	49	B
5	B	10	C	15	A	20	A	25	B	30	C	35	C	40	C	45	B	50	C

7. HOW DO ORGANISM REPRODUCE

SOME IMPORTANT POINTS

- ➢ Reproduction is necessary to continuity on life on Earth.
- ➢ The chromosomes in nucleus contain the information for the inheritance of features for parent to offspring in form of DNA.
- ➢ The basic event is creating the DNA copy
- ➢ DNA copies generated similar not be identical.
- ➢ Variations are thus useful for survival of species.
- ➢ Fission: It is the mode of asexual reproduction.
- ➢ It is two types:
- ➢ Binary Fission: The division of parent cell into identical to daughter cell.
- ➢ Multiple Fission: The division of parent cell into many individual.

- Leis mania reproduce by binary fission. It causes a disease called Kala-azar.
- Fragmentation: The process of breaking the parent cell into fragments called fragmentations. It is also asexual reproduction.
- Regeneration: Planaria and Hydra like animal regenerate its loss body parts in by injury or autonomy.
- The production of new individual from a outgrowth of the parent body due to the repeated cell division at a specific site.
- Reproduction takes place by the vegetative parts of plants (leaves, roots, stem) is called vegetative propagation.
- Spore formation: A spore is a single or a several reproductive celled structure that detaches from the parent and under suitable condition germinates new plants.
- Sexual mode of reproduction.
- The fusion of male and female gametes is called fertilisation.
- In flowers male reproductive part is stamen consists filament, anther and female reproductive part pistil consists stigma, style, and ovary.
- Self pollination: Transformation of pollen grain from anther to flower to stigma of some flower or another flower of same plant.
- Cross pollination: Transformation of pollen grain from anther of flower to stigma of another flower of a plant of same species.
- Fertilization in the human takes place in the fallopian tube.
- Testes produce sperms and male hormone testerone.
- The embryo get nourishment inside the mother body through placenta.
- Sexually transmitted diseases (STD$_s$) are :
- Virus diseases: HIV-AIDS and Warts.
- Bacterial diseases: Syphill and gonorrhoea.

7. HOW DO ORGANISMS REPRODUCE

1.	What is need of reproduction?

	(a)For survival			(b) Continuity of life on earth

	(c)For circulation of blood		(d)None of these

2.	Where are chromosomes present in cell?

	(a)Nucleus	(b)Mitochondria	(c)Cytoplasm	(d)Ribosome

3.	What is the standard form of DNA ?

	(a)Deoxy Nucleic Acid		(b)Deoxyribo Nucleic Acid

	(c)Deoxygenerated Acid		(d)Deoxygenreted Nulcleic Acid

4.	Who is the information source for making proteins ?

(a)Cytoplasm   (b) RNA   (c) DNA   (d)None of these

5.   Which of the following factor leads to variation?

        (a)Temperature   (b)Water levels can vary

        (c)Meteorite hits   (d)All of these

6.   Which of the following mode of the sexual reproduction ?

        (a)Vegitative Propagation   (b)Fission   (c)Spore formation   (d)All of these

7.   Which of the following reproduced by binary fission ?

        (a)Leishmania   (b)Yeast   (c)Both (a)&(b)   (d)Spirogyra

8.   Which disease cause by Leishmania ?

        (a)Cholera   (b)Kala –azar   (c)Hypatits –B   (d)Elphatatis

9.   Which of the following organism reproduce by multiple fission ?

        (a)Sipogyra   (b)Leishmania   (c)Plasmodium   (d)Amoeba

10.   Which of the following of the organism by spore formation ?

        (a)Spirogyra   (b)Leishmania   (c)Plasmodium   (d)Amoeba

11.   Hydra is reproduced by?

        (a)Budding   (b)Regenration   (c)Both (a)&(b)   (d)None of these

12.   Which of the following reproduced by vegetative propagation?

        (a)Rose   (b)Banana   (c)Both (a)&(b)   (d)Papaya

13.   Jasmine is reproduced by?

        (a)Layering   (b)Cutting   (c)Grafting   (d)None of these

14.   Rose in reproduced by ?

        (a)Layering   (b)Cutting   (c)Grafting   (d)None of these

15. Mango is reproduced by ?

    (a)Layering   (b)Cutting   (c)Grafting   (d)None of these

16. In which of the following gamets are formed?

    (a)Fission   (b)Regenration

    (c)Spore formation   (d)Sexual reproduction

17. Rihzopius reproduced by ?

    (a)Budding   (b)Spore formation   (c)Regenration   (d)Fission

18. What is called the process of fusion and male and female gamets ?

    (a)Self pollination   (b)Cross pollination

    (c)Fertilization   (d)Mensuration

19. From which part of plant vegetative propagation takes places ?

    (a)Stem   (b)Root   (c)Leaves   (d)All of these

20. Which of following however is bisexual ?

    (a)Hibscus   (b)Mustard   (c)Both(a)&(b) (d)Watermelon

21. Where fertilization takes place in flowering plant body?

    (a)Fallopian tube (b)Ovary (c)Pollen tube (d)None of these

22. After fertilisation,the zygote divides how many times to form an embryo
    within the ovule?
    (a) one   (b two   (c) three   (d) several times

23. Where fertilization takes place in human beings?

    (a)Vasdeferens (b)Fallopian tube (c)Womb (d)Uterus

24. Which of the following is a party of human male reproductive system ?

    (a)Uterus (b)Cervix (c)Urethra (d)Fallopian tube

25.   Which of the following part of female reproductive system?

(a)Womb (b)Cervix (c)Urethra (d)Both (a)&(b)

26.   Which harmone regulate the production of sperm in males?

(a)Testis (b)Testetrone (c)Exocrine (d)Pancreas

27.   Female germ cell is ?

(a)Sperm (b)Womb (c)Egg (d)None of these

28.   How many eggs are produced every month in females?

(a)1   (b)2   (c)3   (d)4

29.   The time period from fertilization up to birth of the baby is called _____

(a)Germinate period (b)Gestation period

(c)Both (a)&(b)   (d)Gravitation period

30.   Which of the following is sexually transmitted disease?

(a)Warts   (b)Cholera   (c)Pneumonia   (d)Peptic ulcer

31.   Standard form of HIV ?

(a)Humanity immunity virus   (b)Human immunity virus

(c)Human immuno virus   (d)None of these

32.   Standard form of AIDS ?

(a)Inherited manity syndrome   (b)Aquired imuno deficiency syndrome

(c)Aquired imunity definity syndrome   (d)Nonne of these

33.   Which surgical method used in females ?

(a)Vasectomy (b)Sterlization (c)Tubectomy (d)Tuberclosses

34.   Which of the following is bacterial STD ?

(a)Syphills (b)Warts (c)Hiv –Aids (d)Elephantitis

35.   Pistil consist?

(a)Stigma, Style   (b) Ovary, fallopian tube

(c)Stigma, Style, Ovary   (d)Petal

36.   Which of the following in female sex harmone ?

(a)Oestrogen (b)Progestrone (c)Both (a)&(b) (d)Testestrone

37.   What happen when fertilization is not occur?

(a)Implantation (b)Mensuration (c)Plantation (d)None of these

38.   The embryo gets nutrition from mothers blood with help of _____

(a)Implantation (b)Mensuration (c)Placenta (d)Gestation

39.   Mensuration occurs at a regular internal of _____ days .

(a)28   (b)18   (c)20   (d)24

40.   After that the ovaries do not release egg this style is known as _____

(a)Gastation   (b)Menopause   (c)Vascetomy   (d)None of these

41.   which type of reproduction Mucor will reproduced ?

(a)Vegitative Propegation (b)Budding (c)Fission (d)Spore formation

42.   Pollen tube enters the ovule through a small opening called_____

(a)Micropyle (b)Ovary (c)Synergids (d)None of these

43.   How many male gamet have one pollen tube?

(a)1   (b)3   (c)2   (d)Either 3 or 2

44.   In human made reproductive system which part provides an optimal

Temperature is $1\text{-}3^0 c$ lower than normal temperature of the body ?

(a)Vasdeferesns   (b)Testes   (c)Urethra   (d)Scrotum

45.   From which process we can grow seedless plant ?

(a)Pollination   (b)Regeneration

(c)Vegetative propagation   (d)Spore formation

Answers :

Q	A	Q	A	Q	A	Q	A	Q	A
1	B	10	A	19	D	28	A	37	B
2	A	11	C	20	C	29	B	38	C
3	B	12	C	21	B	30	A	39	A
4	C	13	A	22	D	31	C	40	B
5	D	14	B	23	B	32	B	41	D
6	D	15	C	24	C	33	C	42	A
7	A	16	D	25	D	34	A	43	C
8	B	17	B	26	B	35	C	44	D
9	A	18	C	27	C	36	C	45	C

HEREDITY AND EVOLUTION

SOME IMPORTANT POINTS

➤ Inheritance is the result of variation during the process of reproduction.

➤ These variations are essential for survival.

➤ Every individual which are sexually reproduced has two copies of gene. One is dominant and other is recessive.

➤ In offspring of sexual reproduction new combination of traits can be gain.

➤ In different species sex of the offspring is determined by different factors.

➤ In human being sex is determined by the paternal chromosome. Whether it is X (girl) or Y (boys).

➤ Variation can be for only genetic drift or survival advantage or both.

➤ The changes which are due to environment factor on the non reproductive tissue do not affect inheritance.

➤ For a new species genetic drift as well as separation is essential.

- For classifying organism the evolutionary relationship study is done.
- For evolutionary studies fossils are good evidence.
- Complex organs are the result of survival and modification of simple one.
- Different body structures have been adopted and modified as per their functional use and modification in environment.
- Evolution doesn`t signify the progress from simple organism to complex. Evolution both type of organism flourishes at the same time.
- Human being evolved in Africa and then spread all over the world.

8. HEREDITY AND EVOLUTION

1.	The number of successful variation are maximised by the process of?

	a. Sexual reproduction		b. Asexual reproduction

	c. Both (a) & (b)		d. None of these

2.	Inheritance from the previous generation provides both a common basic?

	a. Body design	b. Organ design	c. Mental design	d. None of these

3.	If one bacterium divides and then the resultant two bacteria divide again ,the four individual bacteria generated would be very similar. There would be only very minor differences between them , generated due to only very minor differences between them, generated due to small in accuracies in _____ copying?

a. RNA   b. ATP   c. DNA   d. None of these

4.   All these variation in a species have equal changes of surviving in the

environment?

a. Depending upon on the nature of variation   b. Due to RNA

c. Both (a) & (b)   d. None of these

5.   Bacteria that can with stand heat will survive better in a?

a. Sound wave   b. Heat wave   c. Light wave   d. All of these

6.   Ghanshyam is a male who`s son ear is same as ghanshyam .It is due to

Inherited traits ? choose the correct statement :

a. Ghanshyam and his grand son looks same

b. Ghanshyam son`s is looks same as friend of grand son

c. Both (a)&(b)

d. None of these

7.   Each traits can be influenced by both parental and?

a. Maternal DNA   b. Uncle`s aunty DNA   c. RNA   d. None of these

8.   A mendelian experiment consisted of breeding tall pea plants bearing violet

flowers with short pea plants bearing white flowers . The prognancy all bore

violet flowers but almost half of them were short this suggested that the genetic

make up of the tall parent can be depicted as ?

a. TTWW   b.TTWw   c. TtWw   d. TtWw

9.   Mendel used a number of contrasting visible characters of garden pears?

a. Round/Wrinkled seeds   b. Tall / Short plants

c. White /Violet flowers   d. All of these

10.   The information source for making proteins in the cells is?

a. Cellular DNA   b. RNA   c. Nucleus   d. ATP

11.   A section of DNA the provides information for one proteins is caused?

a. DNA   b. Nucleus   c. Vacuole   d. Gene

12.   Plant height can thus depends on the amount of a particular plant?

a. Hormone   b. Cell   c. Tissue   d. All of these

13.   If the gene for that enzyme has an alteration that makes the enzyme?

a. Less efficient   b.Gain efficient   c. Both (a) & (b)   d. None of these

14.   Due to enzyme less efficient less the plant will?

a. Long   b. Tall   c. Short   d. All of these

15.   In the experiment of Mendel each pea plant must have _____ sets of genes

a. Three   b. Two   c. Four   d. Paired

16.   In Mendelium experiment for this mechanism to work , germ cell must have ____

Gene set ?

a. Million   b. Two   c. One   d. None of these

17.   The fact that each gene set is present not as a single long thread of DNA as

Separate independent pieces each equal a?

a. Chromosome   b. Genes   c. Bacteria   d. DNA

18.   Choose the correct statement :

a. Each cell will have two copies of each chromosome

b. DNA work same as RNA

c. Plant will short due to less enzyme

d. Both (a)&(b)

19.   _____ sexes participants in sexual reproduction.

a. One   b. Three   c. Two   d. Four

20.   _____ can change sex .

a. Snakes   b. Bees   c. Seals   d. Snails

21.   In some type of animal`s eggs _____ decide.

What type of sex born male or female?

a. Atmosphere   b. Temperature   c. Egg layer   d. None of these

22.   Which of the living beings, the sex of the individuals is layerly genetically

Determined ?

a. Insects   b. Mammals   c. Human   d. None of these

23.   _____ inherited from our parents decide whether we will be boys or girls .

a. Chromosomes   b. Genes   c. Cells   d. Cell wall

24.   Women have perfect pair of sex chromosomes caused ?

a. XX   b. XY   c. YX   d. All of these

25.   Man`s sex chromosomes is?

a. XY   b. XX   c. XZ   d. None of these

26.   When X chromosome is shared by mother the born baby is?

a. Girl   b. Boy   c. Dog   d. Cat

27.   There is an un built tendency to variation during reproduction, both because

Of error in?

a. DNA   b. RNA   c. Ceus   d. Bacteria

28.   Evolution depends upon?

        A.Naturaly selected   b. Accidental selected c. None of these d. Both (a)&(b)

29.   The causes of beetle `s decreasing in population is?

        a. Change in the DNA of germ cells   b. Because of starvation

        c. Heavy rain   d. None of these

30.   A cross between two plants having two pairs of contrasting is caused?

        a. Monohybrid   b. Dihybrid   c. Sexualbrid   d. All of these

31.   In human beings there are pairs of chromosomes?

        a. 23   b. 22   c. 24   d. 20

32.   Out of these 22 chromosomes pairs called?

        a. Chromosome   b. Cells   c. Autosomes   d. All of these

33.   The last pair of chromosomes that help in deciding gender of that individual is

        Called?

        a. Genes   b. Nucleus   c. Chromosomes   d. All of these

34.   XX chromosomes present in?

        a. Girls   b. Boys   c. Trasents   d. None of these

35.   Branch of science that deals in heredity and variation called?

        a. Heredity   b. Variation   c. Genetics   d. Physics

36.   The transmission of features /characters traits from one generation to the

        next?

        a. Genetics   b. Heredity   c. Variation   d. None of these

37.   The differences among the individuals of a species /population called?

a. Varaition   b. Heredity   c. Development   d. All of these

38.   The scientist who stated his experiment on plant breeding and hybridisation

Was named?

a. Albert einstien b. George Michel c. Gregor Johnan Mendel d. None of these

39.   The organ that have different origin and structure plan but some function

called ?

a. Organ system   b. Analogous system   c. Fossils   d. None of these

Answers :

Q	A	Q	A	Q	A	Q	A	Q	A	Q	A	Q	A	Q	A
1	A	6	A	11	D	16	B	21	B	26	A	31	A	36	B
2	A	7	A	12	A	17	A	22	C	27	A	32	C	37	A
3	C	8	D	13	A	18	D	23	B	28	D	33	C	38	C

| 4 | A | 9 | D | 14 | C | 19 | C | 24 | A | 29 | B | 34 | A | 39 | B |
| 5 | A | 10 | A | 15 | B | 20 | D | 25 | A | 30 | B | 35 | C | | |

9. OUR ENVIRONMENT

SOME IMPORTANT POINTS

- ➢ Biodegradable substances can be broken down by the action of bacteria's.
- ➢ The non biodegradable substance present in environment for a long time.
- ➢ The ecosystem forms by the all interacting organisms in an area.
- ➢ An ecosystem include both biotic and abiotic components.
- ➢ The organisms which makes our food by inorganic substances are called produces.
- ➢ The organisms which are depend on produces for their food are called consumers.
- ➢ The organisms are which break down the food remains of organisms are called decomposes.
- ➢ The levels of a food chain are called trophic levels.
- ➢ The food chain consists series of organisms feeding on one another with various trophic levels.

➢ The organisms which are the at first trophic levels are produces.
➢ The oraginsms which are at second trophic level are primary consumers.
➢ The secondary consumers form the third and tertiary consumers from fourth trophic level.
➢ The produces capture about 1% of sunlight energy to make its food.
➢ The 10% of energy taken as the average value at each step from producers to the next levels of consumers.
➢ The transfers of harmful chemicals at each trophic level is called biological magnification.
➢ When the three atoms of oxygen combined they form a molecule of ozone O_3.
➢ The ozone layer is not act as an protective blanket and protect us from UV radiation from the sun.
➢ The increasing in uses of CFCs from refrigerator resulted in ozone layer.

9. OUR ENVIRONMENT

1.   A natural phenomenon that becomes harmful due to pollution is?

   a. Global warming   b.Ecological balance   c.Green house effect

2.   The pollutant responsible for ozone holes is ?

   a.CO_2 b.SO_2 c.CO   d.CFC

3.   One of the best solutions to get rid of non biodegradable wastes is?

   a.Burning   b.Dumping   c.Burying   d.Recycling

4.   Animal dung is _____ waste.

a. Biodegradable   b. Hazardous   c. None biodegradable   d. Toxic

5.   Which of the following is biodegradable?

     a. Iron   b. Plastic   c. Lether belts   d. Silver

6.   Which of the following is non biodegradable ?

     a.Animal bones   b.Nylon   c.Tea leaves   d.Wool

7.   Name one non biodegradable waste which may pollute the earth to dangerous levels of toxicity ,if not handled properly ?

     a.DDT   b.CFC   c.PAN   d.Radioactive substances

8.   In lake polluted with plastisides ,which one of the following will contain the maximum amount of pesticides ?

     a. Small fish   b. Big fish   c. Water birds   d. Microscopic animals

9.   The major pollutant from automobile exhaust is ?

     a. NO   b.CO   c.SO_2 d. Soot

10.   The green house gasses,otherwise called radioactivitely active gases includes:

     a. Carbon   b.CH_4 c.N_2O d. All of these

11.   Algal bloom results in ?

     a. Global Warming   b. Salination

     c.Eutrophication   d.Biomagnification

12.   A high biological oxygen demand indicates that ?

     a.Water is pure   b.High level of microbial pollution   c.Absence of microbial

13.   The effects of radioactive pollutants depends upon ?

     a.Rate of diffusion   b.Enegry releasing capacity

c.Rate of deposition of the contaminant   d.All of these

14. The range of normal human hearing is in the range of ?

a.10Hz to 80Hz   b.50Hz to 80Hz   c.50Hz to 15000Hz d.15000Hz to above

15. The pollution which does not persistent harm to life supporting system is ?

a.Noise pollution   b.Radiation pollution

c.Organochlorine pollution   d.All of these

16. Soap and detergents are the source of organic pollutants like ?

a.Glycerol   b.Polyphosphates   c.All of these

17. Growing agricultural crops between rows of planted trees is known as ?

a.Social forestry   b.Jhum   c.Taunga system   d.Agroforestry

18. The main atmospheric layer near the surface of earth is ?

a.Troposphere   b.Mesphore   c.Ionosphere   d.Stratosphere

19. Name of substance whose accumulation in pelicians of lake Michigan led to

the formation of thin shells of their eggs ?

a.CFC   b.PAN   c.DDT   d.PAC

20. Name the process in which a harmful chemical enters the food chain and gets

concentrated at each level in the food chain ?

a.Concentration   b.Expansion   c.Pollution   d.Biomagnification

21. Specific enzymes are needed for the break down of a?

a.Particular substance   b.Solid substance   d.None of these

22. Substances that are not break down in this manner are said to be ?

a.biodegradable   b.Non biodegradable

c.(a) and (b)   d.None of these

23. All organisms such as plants ,animals , microorganisms and human beings

maintain a balance in ?

a.Nature   b.Chain   c.None of these

24. Abiotic components comprising physical factors like ?

a. Temperature   b.Rain fall   c.Soil   d.All of these

25. All living organisms interact with ?

a.Each other   b.Together   c.Own   d.None of these

26. While gardens and crop fields are ?

a. Human made ecosystem   b.Artifical ecosystem

c. Both   d.None of these

27. The radiant energy of the sun in the presence of ?

a. Chlorophyll   b.Photosythesis   c.None of these

28. All green plants and certain blue - green algae which can produce food by?

a. Photosythesis   b.Sunlight   c.Water   d.None of these

29. Organisms depend on the ?

a. Food   b.Plant   c.Producer   d.All of these

30. Choose a correct statement ?

a. Tree – Goat – Tiger   b.Tree –Tiger – Lion

c. Tree – Elephant – Tiger   d.None of these

31. Which are the first tropic level ?

   a. Autotrophs , Hetrotrophs   b.Autotrophs , Producers

   c. Plant and animal   d. None of these

32. What is the example of harbivores ?

   a. Goat   b.Lion   c.Tree   d. None of these

33. What is the example of carnivores ?

   a. Lion   b.Tiger   c.Man (human people)   d. All of these

34. Man is the example of ?

   a. Carnivores   b.Omnivores   c.Parasides   d.None of these

35. Lice is the example of ?

   a. Parasites   b.Omnivores   c.Carnivores   d.None of these

36. In organic substances that go into the soil and are used up once more by the ?

   a. Animal   b.Plant   c.(a)&(b)   d.None of these

37. The food we eat acts as a fuel to provides us energy to do ?

   a.Flight   b.Work   c.Sleep   d. All of these

38. The autotrophs Capture the energy the energy present in sunlight convert it to?

   a. Physical energy   b.Heat energy   c.Chemical energy   d. None of these

39. Chemical energy Supports all the activities of the ?

   a. Non living world   b.world living   c.Animal Plant   d.Plant

40. Some energy is lost to the environment in forms which can ?

   a. Be used again   b.Not be used again

   c.Not used energy   d.None of these

41.   Green plants are eaten by ?

         a. Teritary consumers   b.Secondary consumers

         c. Primary consumers   d.Producers

42.   Human beings occupy which level in any food chain ?

         a. Low level   b.Medium level   c.Top level   d.No level

43.   Ozone is a molecule formed by three atoms of ?

         a.Carbon   b.Helium   c.Oxygen   d. Hydrogen

44.   Ozone is a ?

         a.Deadly poision   b. Layer   c. Harmful   d.None of these

45.   Which is highly damaging to organisms ?

         a.Ultravoilet radiation   b.Animal   c.Omnivores   d. Carnivores

46.   Which is known to cause skin cancerin human beings ?

         a. Sunlight   b.Ultraviolet radiation   c.vitamin   d. None of these

47.   The amount of ozone in the atmosphere began to drop sharply in the ?

         a.1980s   b.1990s   c.1970s   d.None of these

Answers:

Q	A	Q	A	Q	A	Q	A	Q	A
1	C	11	C	21	A	31	B	41	C

2	D	12	B	22	B	32	A	42	C
3	D	13	D	23	A	33	D	43	C
4	A	14	C	24	D	34	C	44	A
5	C	15	A	25	A	35	A	45	A
6	B	16	C	26	C	36	B	46	B
7	D	17	C	27	A	37	D	47	A
8	C	18	A	28	A	38	C		
9	B	19	C	29	C	39	B		
10	D	20	D	30	A	40	B		

10. NATURAL RESOURCES

SOME IMPORTANT POINTS

➢ Life exists on the earth because of resources like soil,water,air and the energy we get from the sun.

➢ The outer crust of the earth is called Lithosphere.

➢ Air is the mixture of many gases like nitrogen,oxygen,carbon dioxide and water vapours etc.

➢ Carbon dioxide is used by two ways;

1. Green plants convert carbon dioxide into glucose in the presence of sunlight.

2. Some marine animals use carbonates dissolve in sea water to make their shells.

➢ Uneven heating of air over land and water causes wind.

➢ Evaporation of water from water bodies and then condensation give us rain.

- Rainfall patterns depend on the patterns of wind.
- An increase of harmful substances in the air or environment is known as air pollution.
- The present source of fresh water is Rain.
- The improvement of harmful substances in the water is known as water pollution.
- The roots of the plants play an important role in preventing soil erosion.
- Nitrogen is an important compound for the synthesis of proteins,vitamins,DNA and RNA etc.
- The endoskeletons and exoskeletons of various animals are also formed from carbonate salts.
- Carbon-dioxide,Methane and Sulphur areknown as the green house gases.

10. NATURAL RESOURCES

1.   The best requirements of all life forms on the earth are fulfilled by?

    (a) The resources available on earth

    (b) Energy from the sun

    (c) Atmosphere present on the earth

    (d) all of the above

2.   The outer crust on the earth is ?

    (a) Lithosphere   (b) stratosphere   c. biosphere   d. None

3.   What part of earth surface covered from water?

    (a) 25%   (b) 75%   c. 80%   d. None

4.   The biotic component of biosphere is?

(a) living   (b) non living   c. both   d. None

5.   Living things are found where exists:

        a. atmosphere   b. hydrosphere   c. lithosphere d. None

6.   The air that covers the earth as a blanket is:

        a. hydrosphere b. atmosphere   c. biosphere   d. none of these

7.   The composition of air due to which life exist on earth is?

        a. oxygen   b. carbon dioxide   c. nitrogen   d. monoxide

8.   The cell who need oxygen to break down glucose molecule and

        get energy are?

        a. prokaryotic cell b. eukaryotic cell c. both   d. none of these

9.   Life supporting of the earth where atmospheres hydrosphere and

        Lithosphere interact is?

        (a)biosphere   b. stratosphere   c. none of these d. both a & b.

10.   The direction of wind during the day would be from?

        a. the sea of the land   b. the land of the sea

        c. both   d. none of these

11.   Winds are created due to the movement of?

        a. air   b. water   c. both   d. none of these

12.   Patterns of rain fall are decided by the prevailing of?

        a. wind pattern   b. forest pattern   c. cloud pattern   d. both a & c

13.   The mainly oxides which cause air pollution on earth are?

a. sodium   b. sulphur   c. nitrogen   d. both b & c

14.   When the harmful gases dissolve in rain its cause?

     a. acid rain   b. tsunami   c. both   d. none

15.   The visible indication of air pollution is?

     a. smog   b. acid rain   c. both   d. none

16.   Fresh water are available in form of?

     (a)ice –caps   (b)underground water

     (c)both (a)and (b)   (d)none of these

17.   What factors causes water pollution on the earth?

     (a)dissolving of fertilizers and pesticides in water

     (b) sewage from our towns

     (c) all of the above

18.   The water that is found inside the deep reservoir would be?

     (a)cold (b)hot (c)cold than upper part (d)none of these

19.   Which factors helps in making of soil?

     (a) the sun (b)water (c)wind (d)all of above

20.   Some bits of decayed living organisms are also found in the soil are?

     (a)humus (b)minerals (c)nutrients (d)none of these

21.   Which factor determines that which plants are grown on soil?

     (a) Nutrients of soil   (b) humus present in soil

     (c) depth of the soil   (d)all of these

22.   The topmost lawyer of soil is?

(a)top soil   (b)medium soil (c)none of these (d)both (a)and (b)

23.   Which organism helps in making rich humus?

        (a)earthworms (b)bacteria (c)fungi (d)none of these

24.   Which factors are responsible for the removal of fine particles from
        soil?

         (a)flowing of water   (b)wind (c)both (a)and (b) (c)none of these

25.   The nitrogen fixing bacteria are most commonly used in the roots of?

        (a) Legumes   (b) carrot (c) photo python (d) none of these

26.   Nitrates and nitrites are converted into amino acids to make?

        (a)proteins (b)carbohydrates (c)fats (d)none of these

27.   The combined state of carbon is?

        (a) Carbon dioxide   (b) carbonates

         (c) Hydrogen –carbonate salts   (d)all of these

28.   The green house gases which lead to global warning are?

        (a) Carbon dioxide   (b) methane (c) both (a) and (b) (d) oxygen

29.   The oxygen from the atmosphere can be used by?

        (a) Respiration (b) combustion (c) formation of oxides (d) all of above

30.   The formulae of ozone is?

        (a)O_2 (b) O   (c) O3   (d) none of these

31.   A hole in the ozone layer was first seen above?

        (a)Antarctica   (b) ammonia (c) South Africa (d) none of these

32.   Ozone layer gets depleted by the used of?

(a)CFCS   (b) hydrogen gas   (c) carbon dioxide gas (d) none of these

33.   The ozone layer protects us from?

(a)U.V rays   (b) harmful gases   (c) tsunami   (d) none of these

34.   Nitrogen is contained in?

(a)nucleic acids   (b)amino acids   (c)fats   (d)none of these

35.   In elemental form carbon access as?

(a)diamonds   (b)graphite   (c)both (a)and (b) (c)none of these

ANSWERS:

QUES.	ANS.	QUES.	ANS.	QUES.	ANS.	QUES.	ANS.
1	d	11	a	21	D	31	A
2	a	12	d	22	A	32	A
3	b	13	d	23	A	33	A
4	a	14	a	24	C	34	A
5	d	15	C	25	A	35	C
6	b	16	C	26	A		
7	a	17	D	27	D		
8	C	18	C	28	C		
9	A	19	D	29	D		
10	a	20	a	30	C		

11. MANAGEMENT OF NATURAL RESOURCES

SOME IMPORTANT POINTS

- We need to manage our natural resources by sustainable methods.
- The demand for all resources is increasing at a expanded rate.
- The management of resources helps not to be exploited and live last for long time.
- There are many things we can do to manage our natural resources out of them three R s major practice. (Reduce, Reuse, Recycle).
- The sustainable method to protect forest resources should be taken in the favour of stakeholders.
- Stakeholders are those people who live near the forest and who are totally dependent on forest resources.
- We need to use the fossil fuels coal and petroleum in very large rare cases, because they will ultimately exhaust.
- Watershed management not only increases the production and income of the watershed community but also mitigates droughts and floods and increases the life of downstream dams and reservoirs.
- We need to reduce our requirements, so that the benefits of development reach everyone for now and for all generations to cone.

11. MANAGEMENT OF NATURAL RESOURCES

1. Which one of the following is an example of biotic component of environment?

(a)Wind   (b)Water   (c)Vegetation   (d)Temperature

2.   Which of the following is an non renewable resource?

(a)Solar energy   (b)Hydrocarbon fuel   (c)Flora and Fauna (d)Nuclear

3.   Sanctuaries are established to?

(a)Conduct ecotourism on wildlife   (b)Protect animals (c)None of these

4.   Global warming has resulted due to ?

(a)Lack of rainfall wlorldwide (b)Increased emissions of CO_2 form automobiles

(c)Oxides of sulphur and nitrogen   (d)None of these

5.   The main source of water in India is?

(a)Rain water   (b) Ground water (c)Surface water   (d)Sea water

6.   Floods are caused by ?

(a)Afforestation   (b)Cutting the forests

(c)Tilling the land   (d)None of these

7.   The ganga runs from gangotri through a hundred towns and cities ?

(a)U.P   (b)U.P and Bihar

(c)U.P ,Bihar,W.B   (d)U.P,W.B,Haryana

8.   Water pollution can be identified by testing its ?

(a)PH level   (b)Biological oxygen demand

 (c)Both   (d)None of these

9.   The three R'S to save the environment are ?

(a)Reserve,Reduce ,Recycle (b)Reuse ,Reserve,Reduce

(c)Reserve,Reuse,Reduce   (d)Reduce,Rcycle ,

10. Which is in Rajasthan?

(a)Surangams   (b)kattas   (c) Kulks   (d)Nadis

11. Kattas is the ancient method of water harvesting found in?

(a)Himzachal pardesh   (b)Karnatka

(c)Tamil nadu   (d)M.P

12. Amirata devi bishnoi sacrificed her life to the protection of the ?

(a)Sal trees   (b)Pine trees

(c) Khejri trees   (d)Alpine meadows

13. In independent india Plantation of which trees caused their monocultures ?

(a)Eucalyptus   (b) Pine

(c)Eucalyptus ,Pine and teak (d)Eucalyptus, Pine, teak and Need

14. The Ganga runs its course from ?

(a)Ganga sagar   (b)Himalays Peak Everest

(c)Gangotri   (d)Jamnutri

15. The presence of which micro organisms in ganga water indicates

Contamination ?

(a)Amoeba   (b)Mucar spores

(c)Coliform bacteria   (d)None of these

16. The Chipko movement started from ?

(a)Reni in gharwal   (b)Arborio forest   (c)Khejrali village

17. Primary source of water is ?

(a)Rivers   (b)Ground water   (c)Lakes   (d)Rain water

18. Tawa irrigation project is in ?

(a)Maharastra (b)Mahdhya pardesh   (c)Orissa   (d)Haryana

19. Measure of biodiversity of an area is?

(a)The number of species found there (b)The range of different life forms

(c)Both (a)&(b)   (d)None of these

20. Which energy of water is used to produce hydro electricity ?

(a)Potential energy   (b) kinetic energy (c) Both (a)&(b)   (d)None of these

21. Chipko Andolan is concerned with?

(a)Conservation of natural resourses

(b)Zoological survey of India

(c)Forest conservation

22. The concept of 'Biosphere Reserves' was evolved by?

(a)Government of India   (b)Botanical survey of India

(c)UNESCO   (d)UNDP

23. Why should be conserve biodiversity because?

(a)We should preserve biodiversity we have inherited

(b)A loos of diversity may lead to aloss of ecological stability

(c)Both(a)&(b)

24. The problems for critism about large dams are that they?

(a)Displace large number of peasants and trebles without proper rehabitation

(b)Contribute enormously to deforestation and the loss the biological diversity

(c)All of the above

25. Natural resourses like?

(a)Soil   (b)Air   (c)Water   (d)All of these

26. Multi-crore project came about in?

(a)1986   (b)1985   (c)1988   (d)1990

27. The Ganga runs its course of over?

(a)2600 Km   (b)2700 Km   (c)2500 Km   (d)2575 km

28. Largely untreated sewageis dumped into the Ganga every?

(a)Day   (b)Month   (c)Year   (d)All of the days

29. The pollutants are?

(a)Useful   (b)Harmful   (c)Both   (d)None of these

30. Will ultimately be exhausted?

(a)Fossil fuels   (b)light   (c)Petroleum   (d) (a)&(c)

31. Petroleum is used?

(a)Car & Bike   (b) Cycle, lamp   (c) None of these

32. Coal and petroleum have been formed from?

(a)Gobar gas   (b)Biomass   (c)None of these

33. Water harvesting is an age – old concept in?

(a)India   (b)Kerala   (c)America   (d)China

34. Parts of Himachal Pradesh had evolved local system of canal irrigation called?

(a)Kulhs   (b)Ponds   (c) Surangams   (d)Eris

35. kulhs is used for?

(a)Irrigation   (b)Drink   (c)None of these

36.   Rains in India are laregely due to the?

(a)Monsoons   (b)Winter   (c)Wind   (d) None of these

37.   If the goals of all the stakeholders with regards to the management of the forest

are?

(a)Medium   (b)Same   (c)High   (d)None of these

38.   What is used to make slats for huts?

(a)Plants   (b)Wood   (c)Bamboo   (d)None of these

39.   Coal is used in?

(a)Thermal power   (b)Chemical idustury   (c)Physical laboratory   (d)None of these

40.   Who was started Chipko Andolan?

(a)Ramlal Bahuguna   (b) Sunder lal Bahuguna

(c)Ramesh lal Bahuguna   (d)Shyam lal Bahuguna.

Answers :

Q	A	Q	A	Q	A	Q	A	Q	A	Q	A	Q	A	Q	A
1	C	6	C	11	B	16	A	21	A	26	B	31	A	36	A
2	B	7	C	12	C	17	C	22	C	27	C	32	B	37	B
3	B	8	C	13	C	18	B	23	C	28	D	33	A	38	C
4	B	9	D	14	C	19	C	24	C	29	B	34	A	39	A
5	A	10	D	15	C	20	A	25	D	30	D	35	A	40	B

SOURCE OF ENERGY

SOME IMPORTANT POINTS

- Energy neither be created or destroy its only courted from one form to another
- A good source of energy would be that have high calorific value be easily accessible be easy to store and transport be eco-friendly
- Energy are characterized into two types
  1. Conventional source of energy
  2. Non-conventional source of energy
- Fossil fuels, thermal power plant, hydro-power plant, bio-mass and wind energy are the examples of conventional source of energy
- Solar energy tidal energy, wave energy, ocean thermal energy, nuclear energy, and geo-thermal energy are the examples of non-conventional source of energy
- Coal petroleum and natural gas the example of now-renewable source of energy
- Air, water, solar ration and tidal energy are the example of renewable source of energy
- hydro power- plants converts the potential energy of thee falling water into electricity by making dams
- animals waster plant waste and other bio degradable wastes used for making bio-mass
- bio-gas is a excellent fuels contains 75% of methane
- to is stalling a wind mill we required a very large area and the minimum speed of wind is 20km/h
- solar cells made up of silicon and its converts light energy into electrical energy
- nuclear energy is divided into two types
  1. Nuclear fission
  2. Nuclear fusion
- the splitting of heavy atom into many lighter atoms to gives energy is known as nuclear fission
- two of more atoms combined to makes a very atom to gives energy

➤ the energy stores as heat inside the earth is used to rotate the turbine and then to the generator converts heat energy into electrical energy is known as geo-thermal energy

12. SOURCES OF ENERGY

1. The chemical energy in the wax converted to heat energy and light energy on?

(a) burning   (b)washing

(c) cooling   (d) none of these

2. Sun's energy is due to nuclear?

(a) fission   (b) fusion   (c) collaps   (d) none of these

3. wood was the comman source of ?

(a) chemical energy   (b) heat energy

(c) electrical energy   (d) muscular energy

4. The growing demand for energy was largely met by the fossil fuels?

(a) coal   (b) petroleum

(c) a & b   (d) none of these

5. Coal and petroleum are?

(a) limited   (b) unlimited   (c) none of these

6. The fossil fuels are which sources of energy?

(a) Renewable   (b) non renewable   (c) none of these

7. Burning fossil fuels has other?

(a) Disadvantages too   (b) advantages too

(c) non advantages   (d) none of these

8. The air pollution caused by burning of which products?

(a) Coal   (b) petroleum

(c) both   (d) none of these

9. ------------- rain which effects our water and soil resources?

(a) acid   (b) base

(c) neutral   (d) none of these

10. Fossil fuels are the major fuels used for?

(a) Heat energy   (b) chemical energy

(c) generating electricity   (d) none of these

11. Charcoal burns without?

(a) heat   (b) flames   (c) none of these

12. The starting material is mainly cow –dung, it is known as?

(a) bio –dung   (b) gobar gas   (c) a & b   (d) none of these

13. Wind energy is an environment friendly and efficient source of?

(a) renewable energy   (b) heat energy   (c) non

renewable energy   (d) none of these

14. The wind speed should also be higher than?
    (a) 15kn/h   (b) 10km/h   (c) 20km/h   (d) 30km/h

15. For a 1Mw generator the farm needs about?
    (a) 2 hectares of land   (b) 3 hectares of land
    (c) 4 hectares of land   (d) none of these

16. The initial cost of establishment of the farm is quite?
    (a) technological   (b) electrical
    (c) mechanical   (d) none of these

17. Which progress our demand for energy increases day by day?
    (a) technical   (b) electrical
    (c) mechanical   (d) none of these

18. Which we use to do more and more of our tasks?
    (a) Machines   (b) hard work

    (c) hand   (d) all of these

19. Our demand for energy?

    (a) increases   (b) decreases

    (c) remains constant   (d) none of these

20. India receives the energy equivalent to more than?

    (a) 5000 million kwh   (b) 5000 billion kwh

    (c) 5000 trillion kwh   (d) none of these

21. The average distance between the sun and earth is called?

    (a) Energy constant   (b) heat constant

    (c) Solar constant   (d) light constant

22. A large number of solar cells are combined in an arrangement of, it is known as?

    (a) solar cell panel   (b) solar system

(c) battery   (c) none of these

23.   Solar energy is?
       (a) limited   (b) unlimited

       (c) constant   (d) none of these

24.   What is the example of solar – cell panel?

       (a) TV   (b) radio

       (c) none of these

25.   The sea – shore can be trapped in a manner to generate?

       (a) Electricity energy   (b) heat

       (c) Energy   (d) none of these

26.   Renewable energy is available in our?

       (a) Natural environment   (b) science laboratory

       (c) none of these

27.   A solar water heater can be used to get hot water on?

       (a) A sunny day   (b) a cloudy day

       (c) A wind day   (d) all day

28.   Many of the sources ultimately derive their energy from the?

       (a) Solar   (b) bio- mass

       (c) Moon   (d) sun

29.   Green house effect is caused by gases like?

       (a) helium   (b) oxygen   (c) nitrogen   (d) CO_2

30.   A solar water heater cannot be used to get hot water on?

(a) A sunny day   (b) a cloudy day

(c) A hot day   (d) a windy day

31.   Chief constituent of natural gas is?

(a) Methane   (b) ethane   (c) butane   (d) propane

32.   The power plant which converts potential energy of falling

Water into electricity is?

(a) Nuclear plant   (b) thermal plant

(c) Hydro plant   (d) wind plant

33.   Wood is a

(a) Primary fuel   (b) secondary fuel

(c) Liquid fuel   (d) processed fuel

34.   Which of the following is not an example of a bio-mass

Energy source?

(a) wood   (b) gobar gas   (c) coal   (d) nuclear

35.   The popular name of biogas is?

(a) gobar gas   (b) marsh gas

(c) ethane gas   (d) helium gas

36.   The country of winds is?

(a) India   (b) china   (c) Denmark   (d) Netherlands

37.   Which of the following is the ultimate source of energy?

(a) water   (b) uranium   (c) sun   (d) fossils fuel

38.   Acid rain happens because?

(a) sun leads to heating of upper layer of atmosphere

(b) Burning of fossil fuels release oxides of carbon nitrogen

and sulphur in the atmosphere

(c) Earth atmosphere contains acids

(d) None of these

39. Which one of the following forms of energy leads to least environmental pollution on the process of its harneshing and utization?

(a) nuclear energy   (b) thermal energy

(c) Solar energy   (d) geothermal energy

40. The major problem in harneshing nuclear energy is how to   (a) split nuclei   (b) sustain the reaction

(c) dispose off spent fuel safely (d) none of these

41. The major (gas) constituent of biogas is?

(a) Methane   (b) carbon dioxide

(c) Hydrogen   (d) hydrogen Sulphide

42. Which part of the solar cooker is responsible for green house effect?

(a) mirror (b) glass   (c) black colour (d) none of these

43. Fuel used in thermal power plant is?

(a) water (b) uranium   (c) biomass (d) fossil fuels

44. Large eco systems are destroyed when submerged under the water?

(a) In dams   (b) on dams

(c) Over the dam   (d) none of these

45. Opposition to the construction of Tehri dam on which river are due to problems?

(a) Yamuna (b) Narmada (c) ganga (d) Brahmaputra

46. Fuels are of which product?

(a) plant (b) animal (c) plant & animal (d) none of these

47. Plant has which like structure built with bricks?
(a) dome   (b) dame   (c) dam   (d) none of these

48. The digestor is a sealed chamber in which there is no?

(a) oxygen   (b) carbon   (c) helium   (d) methane

49. Bio gas is an excellent fuel as it contains up to ----------% methane?

(a) 73%   (b) 75%   (c) 80%   (d) 85%

50. Methane heating capacity is?

(a) Low   (b) high   (c) none of these

51. Bio gas used for?

(a) Lighting (b) plant (c) animal (d) none of these

52. Bio mass is a which source of energy?

(a) Renewable (b) non- renewable (c) heat energy

(d) Chemical energy

53. In a water lifting pump the rotatory motion of windmill is utililsed to lift water from a well is an example of?

(a) Kinetic energy   (b) energy

(c) Heat energy   (d) none of these

54 wind energy is also used to?

(a) Generate electricity   (b) electrical energy

(c) heat energy   (d)chemical energy

55.   A number of windmills are arranged over a large area is

Called

(a) Wind energy   (b) heat energy

(c) Chemical energy   (d) none of these

Answers:

Q.	A.	Q.	A.	Q.	A.	Q.	A.	Q.	A.	Q.	A.
1	A	11	B	21	C	31	A	41	A	51	A
2	B	12	B	22	A	32	C	42	B	52	A
3	B	13	A	23	B	33	A	43	D	53	A
4	C	14	A	24	C	34	D	44	A	54	A
5	A	15	A	25	A	35	A	45	C	55	A
6	B	16	B	26	A	36	C	46	C		
7	A	17	A	27	A	37	C	47	A		
8	C	18	A	28	D	38	B	48	A		
9	A	19	A	29	D	39	C	49	B		

10	C	20	C	30	B	40	C	50	B		

SUMMARY

13. IMPROVEMENTS IN FOOD NUTRIENTS

- Plants too require nutrients to fulfill their growth demands. There are thirteen nutrients which are essential for plants and which are supplied from soil to the plants.
- Macro - nutrients out of thirteen nutrients, six nutrients are required by plants in large quantities, which are called macro – nutrients.
- Micro – nutrients remaining seven nutrients are required in small quantities which are called micro nutrients.
- Manure and fertilizers are one of the main sources of nutrients, supplied to plants.
- The organic farming is a type of farming in which organic manures, recycled farm wastes and bio – agents with healthy cropping systems are used.
- Farming of two types of crops together on a same piece of land is called mixed cropping.
- The growing of two or more types of crops in a row pattern is known as Inter – cropping.
- The farming of different crops on a piece of land in a planned successive pattern is known as crop rotation.
- Animal husbandry is type of complete care for farm animals, such as shelter, feeding, breeding and disease.
- Fishes are obtained from both marine and inland resources.
- The echo sounds and satellites are used to guide the fishing nets to capture marine fish.
- The composite fish culture system is a best place for fish farming.
- The capacity of honey collection is very in Italian bees.
- The Indian bees that are used for the production of honey are apiscerana India.
- Poultry farming is undertaken to raise domestic fowl for egg production and chicken, meat.

13. IMPROVENENT IN FOOD RESOURCES

1. The berseem crop is a type of :-

   (a) rabi season crop   (b) fodder crop

   (c) kharif season crop   (d) none of these

2. Wheat are fall in which category of crop:-

   (a) rabi season crop   (b) fodder crop

   (c) kharif season crop   (d) none of these

3. The period of kharif season crop is starts from:-

   (a) November to july   (b) june to October

   (c) march to December   (d) none of these

4. The crossing between different species to same genus refers to:-

   (a) pasturage   (b) wider adaptability

   (c) Hybridisation   (d) none of these

5. Which type of nutrients plants not get from the soil:

   (a) hydrogen and chlorine

   (b) carbon and oxygen

   (c) nitrogen and phosphorus

   (d) none of these

6. Which category of nutrients related to zinc.

   (a) micro-nutrients   (b) manure nutrients

   (c) macro-nutrients   (d) none of these

7. Which pair of crops not used in mixed cropping :-

(a) wheat-groundnut

(b) pulse-mustard

(c) wheat-maize

(d) none of these

8.   Which cattle breeds are used for increasing lactation period.

(a) red sindhi + sahiwal

(b) jerrey + brown swiss

(c) brown swiss + sajowal

(d) none of these

9.   Which fishes are commonly called as bottom feeder:-

(a) mrigals

(b) rohu

(c) starfish

(d) lobster

10.   Which breed of bee is used for commercial honey production:-

(a) A. dorsata

(b) A. florae

(c) A. meligera

(d) none of these

11.   Which of the following are fresh water fish:-

(a) rohu and mrigals
(b) starfishand lobster

(c) rohu and dphins

(d) none of these

12.   Which is net falls in category of cereal crop

(a) wheat

(b) mustard

(c) maize

(d) none of these

13. Pigeon pea is generally known as:-

    (a) chana   (b) moong

    (c) arhar   (d) masoor

14. Which in not falls in category of oil sed crops:-

    (a) soyabean   (b) lentil

    (c) castor   (d) none of these

15. Which crop is not used as a food for the live stoclc:-

    (a) linseed   (b) oats

    (c) sudan gass   (d) none of these

16. Which is not falls in the category of khasib crops

    (a) maize   (b) paddy

    (c) wheat   (d) none of these

17. Which crop shocus the characteristics of tallness and profuse braching.

    (a) oil   (b) fodder

    (c) cereal   (d) none of these

18. Assel is known as a desi breed of

    (a) cow   (b) sheep

    (c) poultry   (d) none of these

19. Which breed of bee is known as rock bee:-

    (a) apis florae   (b) apis dorsata

    (c) apis serena   (d) none of these

20. The method of bish farming is also known as:-

(a) culture fishery

(b) fish adaptation

(c) capture fishing

(d) none of these

21. Which of the following is a finned fish:-

(a) mullets

(b) mollusks

(c) starfish

(d) none of these

22. The capacity of honey collecting are very high in:-

(a) Indian bees

(b) italian bees

(c) german bees

(d) none of these

23. Which are of the following is Indian breed

(a) assel

(b) leghorn

(c) cerana

(d) none of these

24. Which one of the following is foreign breed of cock

(a) assel

(b) meghorn

(c) cerana

(d) none of these

25. Which is the point where sea water and fresh water mix together

(a) brakish water

(b) salt water

(c) prawn water

(d) none of these

Q.	A.	Q.	A.	Q.	A.	Q.	A.	Q.	A.
1	B	6	A	11	A	16	C	21	A
2	A	7	C	12	B	17	B	22	B
3	B	8	B	13	C	18	C	23	A
4	C	9	A	14	B	19	B	24	B
5	B	10	C	15	A	20	A	25	A

NOTES

www.ingramcontent.com/pod-product-compliance
Lightning Source LLC
Chambersburg PA
CBHW080822180526
45168CB00006B/2547